The Water Harvester

Episodes from the Inspired Life of Zephaniah Phiri

Mary Witoshynsky

Published by Weaver Press, Box A1922, Harare, Zimbabwe
in the year 2000

© Mary Witoshynsky, 2000

Designed, illustrated and typeset by Jane Shepherd.
Cover illustration adapted from a woodcut by Marjorie Wallace.
The quotation on page vii is used with the kind permission of Anna Brazier, author of *The Water Harvesting Innovations of Mr Phiri Maseko*, The Natural Farming Network, Harare, 1995. The author would also like to thank Emily Johnson and the Zvishavane Water Project for making available their 1995 booklet, *Water Harvesting and Soil Conservation*.
Printed by Préci-ex Ltd, Mauritius
Distributed in Europe and the USA by the African Books Collective, Oxford, UK
e-mail: abc@dial.pipex.com
http://www.africanbookscollective.com
Fax: 44 01865 793 298

The publication of *The Water Harvester* has been supported by The Ford Foundation, New York, in recognition of the work of Mr Zephaniah Phiri Maseko. Weaver Press and the author express sincere gratitude for their generous assistance, without which the publication of this book would not have been possible.

All rights reserved. No part of this publication may be reproduced, stored in a retrieval system, or transmitted in any form by any means, electronic, mechanical, photocopying, recording or otherwise, without the prior permission of the copyright holders.

ISBN: 0-7974-2123-8

THE WATER HARVESTER

1	Introduction
9	Part One: Beginnings
27	Part Two: Chimurenga
37	Part Three: A Water Harvester's Gospel
57	Illustrations: Mr Phiri's Water Harvesting

Dedication

With love to my father who taught me to value humanity, to my mother who taught me what is important to document; and to Mr Phiri for sowing the seeds of a new mission.

Acknowledgments

This book attests to the commitment of Weaver Press to raise the voices of people with something of value to communicate but who otherwise might not be as widely heard. Such collaborations richly reflect the broader national will of Zimbabweans to work together to build their country and advance its great potential. As a visitor, I have been privileged to share in this endeavour and it is therefore in honour of this spirit of achievement that the story of Zephaniah Phiri, *The Water Harvester*, is offered to the people of Zimbabwe, and to the world.

As part of my research, I was kindly received for interviews with four public figures who played prominent leadership roles in both colonial Rhodesia and independent Zimbabwe: Ian Smith, Garfield and Grace Todd and Victoria Chitepo. Many more people offered valuable insights, information and documents on the social and agricultural history of Rhodesia and Zimbabwe: Anna B, Joshua D, Margaret M, Margaret N, Kenneth P, Florence W, John W; and, to Jane S, Irene S and Ken W, my deepest gratitude for making this book happen.

The staff of the Zvishavane Water Project (ZWP) were most helpful throughout, as were the many people who welcomed me at each ZWP project site I visited. My special thanks to the members of the Phiri family who have brewed innumerable cups of tea for their thousands of visitors over the decades, and who shared information and enthusiasm. Generous assistance from Ginner C, Morgan K, Canaan M and Charles M, master farmers all, and Márcia Maria S, facilitated both research and writing.

A number of individuals encouraged this work at various stages of development: Charles B, Lucy B, Sebastian C, Jorge F, Diane H, John H, Mary J, Ronny J, Olga K, Amanda M, Deborah MS, James M, Raúl S, Robin S, Salim S; and, a very special group: Katy B, Lady L, Murray M, Meredith N, Gerry W and James Z, contributed significant time to reviewing the manuscript.

To my cherished siblings, John and Amanda, Mike and Ruth, and Tom and Jo, thanks for your support. And to my parents, Alex and Gerry, together with José and Lady, I owe an unrepayable debt of gratitude.

With thanks to my beloved G for bringing us to Harare. And, for your gracious welcome, to the fine people of this alluring land - Zimbabwe - *ndinotenda*.

When one is committed, one never feels tired.

Zephaniah Phiri

Introduction

"Life," Mr Phiri[1] says, "is slippery." His observation springs from long years spent at the margins of survival as a subsistence farmer. Mr Phiri's experience is rooted in Zimbabwe's communal lands, the contemporary offshoots of native reserves originally demarcated by colonial Rhodesia's European settler governors. Land segregation uprooted tens of thousands of African families from their homelands and offered little recourse but to coax sustenance from isolated parcels of unproductive ground scattered across the colony's hinterlands. Ownership and commercial exploitation of productive lands was (with limited exception) henceforth reserved exclusively for Rhodesia's settler farmers, ranchers, miners, merchants and industrialists.

Upon independence under African majority rule in 1980, Rhodesia and its native reserves were reincarnated as Zimbabwe and its communal lands. The plots in large part remained state property, and new name notwithstanding, these lands have yet to be communally farmed. Before long, the young nation enjoyed a significant agricultural production boom facilitated by increased government attention to the needs, as well as the skills, of African small-scale and subsistence farmers. Within a decade, however, generations of rural population growth, recurring drought, and over-emphasis on the transfer of large-scale commercial farming techniques to smallholders had generated a growing crisis both for natural resources and for the millions of families whose welfare depends on them. While water and soil diminished, fiscal limitations rapidly merged with the inherited lack of political will for equitable governance, thereby precluding the

[1] Pronounced **peer**-y

new leadership from meaningful engagement with their public mandate to upgrade rural livelihoods in this nation of farmers. At the turn of the 21st century, the situation persists.

But amidst the tumult of settler occupation and protracted civil war that described the colony from 1890 to 1980, some individuals adeptly improvised coping mechanisms to ease the emotional and physical stresses born of the extreme social inequities that were Rhodesia's hallmark. As a mid-twentieth century native reserve farmer, Mr Phiri's land allocation was not well endowed for crop production, being situated in a severely drought-prone area of poor soil quality. Yet through his enlightened vision, Mr Phiri transformed a seemingly exhausted fossil stream into a thriving natural resource on which he would not only build a 'water plantation', but also a mission dedicated to community well-being, the Zvishavane Water Project (ZWP). Such waterways as Mr Phiri nurtured, however, along with marshes, bogs and rivers, constitute wetlands that had been declared off limits to farming operations, and he was made to pay a price for his civil disobedience long before he gained recognition as a gifted innovator.

Where fragile wetlands are commercially cultivated by mechanized ploughing and the use of chemical inputs, their natural fertility withers within a handful of harvests. Such techniques were introduced in early settler days when wetlands could be farmed into virtual extinction, their water depleted and soil nutrients exhausted. Subsequently, legislation to control resource degradation restricted wetlands agriculture. With the loss of access to marshy wetlands, subsistence farmers turned increasingly to stream-bank cultivation, a practice that undermines the health of watercourses due to siltation and, in some places, eventual dehydration. Because wetlands deteriorate so rapidly from invasive agricultural regimens, environmental protection laws banned cultivation within thirty metres of all natural water sources. For the thousands of farmers who had long been denied fair access to land, the vote, to capital, to infrastructure and now to water, these measures imposed new social, political and economic restrictions on the well-being of indigenous populations throughout Rhodesia.

Historically, however, when peoples of pre-colonial Africa were free to traverse the landscape, and when agricultural production was guided by community requirements rather than by commercial interests, wetlands cultivation was a traditional livelihood mechanism. The moist, nutrient-rich soil was not ploughed, seeds were sown by hand, and machinery and chemicals were unknown. With broad and unbordered pastoral expanses, wetlands, indeed croplands and grazing lands in general, were left to lie fallow and rejuvenate long before their life-giving properties were laid to waste. With the wetlands farming ban, the colonial obliteration of ancient custom struck another dissonant chord that reverberated throughout Rhodesia's African heartland.

Whereas many controls on African production and lifestyles were revoked upon independence, some remain, including the wetlands farming regulations, although today these are less strictly enforced. Balanced compromise on this dilemma of resources vs. livelihoods may never be reached, yet the implementation of ecologically sound farming techniques combined with wetlands protection is vital for Zimbabwe, as for all countries. The agricultural conservation methods that Mr Phiri has pioneered offer a means to address these matters in resource-poor zones.

Acting on his own in the 1960s, Mr Phiri took it upon himself to tackle the disparities of his livelihood situation in order to ensure the survival of his family. He envisioned a process that would enable him not only to gently cultivate, but to simultaneously rejuvenate the fossil stream and make it live once more. Revisiting the wisdom of the ancients while being mindful of modern conditions, Mr Phiri's careful experimentation and research revealed the promise of renewed vitality to the natural resources on his small patch of Earth.

Although he suffered numerous consequences for developing these practices under the colonial régime, Mr Phiri's innovations today contribute to the growing field of investigation into viable resource management options for subsistence and small-scale farmers across sub-Saharan Africa. But perhaps most importantly, a major impetus for Mr Phiri in spreading his knowledge via the ZWP is his awareness, learned from a lifetime of experience, that African farmers have had little choice but to be increasingly responsible for helping one another, for outside assistance rarely makes its way to the places where it is most crucially needed.

Numerous articles and studies have been written about the water and soil conservation techniques that Mr Phiri has innovated over the course of his lifetime, and for which he has earned broad recognition. *The Water Harvester*, however, is not one of these volumes. It is the personal archive of the man himself, expressed in his own words. It is not a rags to riches story, for the Phiri family remains on the same small plot, located in a region where paved roads, electricity and telephones are still dreams for the future. It is instead the story of a wealthy spirit, of a man who cultivated an Eden, not for himself, but for the dry-land farmers of the world.

Because I am not a farmer or a conservationist, Mr Phiri's experience engendered in me another appreciation for the significance of his lifework, a life of diligence and devotion, of dignity and caring, of reflection and innovation, and of respect for people and for the planet we share. All this, with a generous portion of humour, too. His quiet heroism couples the denial of victimization with an unflappable determination to uphold human dignity against all odds. For, rather than allow himself to be broken by the numerous injustices of apartheid inflicted directly upon his personal existence, Mr Phiri instead steeled his will to dedicate his life to the service of our humanity. Since knowing him, I no longer enumerate reasons why I can't. I now ask: Why can't I?

Mr Phiri spreads his water harvester's gospel through the mission he founded, the Zvishavane Water Project, an indigenous non-governmental organization (NGO) based on his innovations in self-reliance through natural resource conservation, that operates in southern Zimbabwe. Although the Bible provided the immediate inspiration for his research and experiments, Christian evangelism is not an objective of his conservation agenda. "Christians? No, we don't work with them!" was Mr Phiri's humorous response to a suggestion that he devote ZWP energies to church building rather than water harvesting. "When I look in churches, I see only Christians. Ah, but when I look around here," he said, waving his arm gently across the scene, "I see the people, and we can work with the people!" And so he has worked, for three decades and counting. Beyond this, Mr Phiri addresses agricultural extension and research teams throughout Africa, as well as observers, development professionals and fellow innovators from around the world.

This book conveys the driving forces behind Mr Phiri's philosophy and his many achievements. Along the way, it presents a slice of life in Zimbabwe's communal farm areas as well as a glimpse of life under the colonial system that governed the country until full independence in 1980. Based on personal interviews conducted with Mr Phiri in 1995-96, the text is complemented with information gathered through research and further interviews with former government officials, colleagues and friends of Mr Phiri, and with numerous Zimbabweans in Zvishavane, Harare and Bulawayo, as well as at ZWP community project sites.

Divided into three sections, in Part One: Beginnings, Mr Phiri shares scenes from his childhood and dissects the Bible passages that sparked his renowned innovations. His work was periodically derailed by the Rhodesian régime and further complicated by the psychological and physical brutality of war. In Part Two: Chimurenga, Mr Phiri relates his war-time experience and sheds light on the senseless cruelty that shrouded the lives of communal area villagers throughout the region. In Part Three: A Water Harvester's Gospel, Mr Phiri outlines the post-independence circumstances that led to his founding of the Zvishavane Water Project, a grassroots organization that today is well known and respected far beyond Zimbabwe's borders. Mr Phiri tells of the many ways in which he 'got the chance' to improve his life, and that of ZWP's broad community of farm families.

Descriptions of his 'water plantation' also offer insight into techniques, that when properly implemented and monitored, may well survive the test of time in rehabilitating the wounds of injured wetlands and revitalizing them for use by generations to come. Mr Phiri's hand-made constructions and excavations - Phiri Pits - offer rainwater a permanent resting place from which to safely resist the debilitating consequences of uncontrolled waste through run-off and erosion. Phiri Pits harness both soil and rain where farmers need them most: in their grain and vegetable fields.

Beyond this immediate agricultural benefit, however, the broader potential of soil and rain protection on drought-plagued terrain offers the long-term promise of revitalizing entire zones. For the peoples of impoverished rural areas around the Earth, water sourcing is a growing obstacle to survival. In so many villages, when the water in a stream or a well is gone, it has gone completely; there is no place else to get it. It has ceased to exist. Mr Phiri's innovations show the world that this no longer need be so.

This book has been written for communal area farmers and commercial farmers, shopkeepers and dentists, homemakers and shoemakers, conservationists and scientists, mechanics and artists, students and development professionals, globetrotters and armchair travellers, for leaders and for followers, for Zimbabweans and for everyone. Yet because effective resource management practices are specific to each area where they are implemented, this is not intended as a handbook. Rather, it is a tribute to Mr Phiri and as such seeks to accomplish something more than simply make his contributions more widely known. It seeks to kindle two topics of consideration: Is poverty found only in an empty pocket, or also in the emptiness of hearts and minds? What is there to keep anyone among us from contributing more generously to the well-being of our collective humanity and the Earth we share? If no answers spring readily to mind, may Mr Phiri's words and deeds offer inspiration to follow.

MW
Rio de Janeiro
January 2000

 Part One **Beginnings**

My name is Zephaniah Phiri Maseko. I was born here, in Zvishavane, Zimbabwe, in 1927. Over the years, thousands of people have come to visit my farm and my pond. They come to see my water harvesting. When people visit, they also want to know about me. This is what I tell my visitors.

I feel God created a human being and Earth the same because, in me, there is blood circulation. And yet in the soil, water also circulates. When you dig a deep hole into the soil, God's nature is going to send this circulation into that well, that hole. And the healing of that hole is mostly done by water. Gradually, water seeps into that broken area and then fills up. If you look at it properly, you will find that, within the system, some small elements can be seen. Very fine soil and some small stones fall within that pit.

To me, that shows it's a living system. Just as it happens to me if I break – if I get myself cut. My blood oozes through the broken part, then it clots. After the clotting, then is the healing, because God really does not want me to have a part of my body broken. Just in the same way, the soil. So, through that way, in my land I invite water to come for healing.

Yet it is also a chance for Zephaniah to have water for watering his vegetables, his trees, whatever. My fish enjoy that water.

I began to harvest water in the 1960s. It was after I had been a detainee in Gonakudzingwa detention camp. Before then, I had worked with the National Railways. At that time, it was called Rhodesian Railways, because that was long before our independence. During those years, the spirit of liberation came in and I got myself involved in many political issues. I had quite a lot of quarrels over the government administration of our land issues. I started asking why we were not allowed to farm within the wet areas. Then all this involved me into quite a strong misunderstanding and the reasons why I was being arrested were not proved to me. But I did not give up.

I had joined an African railway workers' union, and one day, while we were having one of our meetings, the police came and rounded us all up. After being arrested for all these politics, I was taken to Gonakudzingwa detention camp and so I slept there for two years. It was not a good place, surely, but it was also not a horrible place.

My fellow detainees and I, we sat around and got fat, they fed us so much sadza. Sadza is our national dish. It is maize-meal porridge. At home, when we are lucky, we eat it together with some nice relishes. Maybe some vegetables and some chicken, maybe some goat stew. You see we all grow our own maize around here, and our own chickens and vegetables. But sometimes the rains are bad and our harvests are very small. So that is why I say, in Gonakudzingwa detention camp, we all got fat. We ate a lot of sadza because, as their prisoners, the government had to feed us. Our wives and our children back home at our farms, they could only eat what we had stored from the last year's harvest.

Now, the settler government liked everybody in the world to think that we Africans in Rhodesia were happy that the settlers came here. Ian Smith was the prime minister back in those days. A friend of mine visited him not long ago. He told her that his one wish in life had always been to make Zimbabwe a better country for all people to live in. But I asked my friend, I said, "Did he really mean *all* the people?"

Anyway, they liked to keep us so quiet that we couldn't even have a peaceful meeting amongst ourselves. Court trials failed to put a blame on many people, but they were taken by force to Gonakudzingwa anyway. They just took people to stand them out, to separate them from their families, so that you begin thinking, "Now look here – my family is *there*!" You come up to say, "Well, no, I don't want politics any more." Yah, but people resisted all this, so that's how I happened

to be at Gonakudzingwa. We had broken no laws, but there we went.

When I came home from Gonakudzingwa, I was told by the Rhodesian Railways that I had long been dismissed from work. And the Native Commissioner told me that I was blacklisted – I wouldn't have any employment by a white man in Rhodesia. This was a big problem, sure. I was already the father of six children, plus myself and my wife: Eight! Hardships. So I began to think, "How was I going to survive with my wife and six children?" That's when I discovered.

One day I was reading from a book, from the Bible, Genesis, Chapter II, where Adam was offered a garden by God. The secret that touched me then was these rivers that ran across the Garden of Eden.

Now, to my knowledge, Adam did not know how to plant a tree or a vegetable. Or to use the water in the river; to take the water and water the trees there. He didn't know about it. But God made it that Adam survived from the fruits in the garden.

So when I thought about the Tigris River, the Euphrates River, I picked up that the sole survival of these trees and Adam was because the rivers had water. Nature – God – made it that these trees survive from the moisture, from the seepage that came from the rivers. So it touched me.

I then thought, "How can I also have this kind of water since in my area there is no river?" Then I started. You know, when the rains fall, there is water pouring, the run-off water, from the *ruware* near my house. So behind my house, I made a sand trap to catch the run-off soil and water. And I could watch water stopped by the trap. Then after some two, three weeks after the rain, I could notice the crops near the sand trap still survived, even though my other crops were getting drought: Dry!

Then I saw that this water was life.

I started making all these sand traps around here. I started a reservoir. I watched the water that fell from the roof of my house. Then I made a small tank so that I could harness the run-off from my roof. So all these were the ideas of getting survival for my family. When I did this, it really jerked me up, because I got the chance. When other peoples' crops here failed, my crops always proved a success, because of my water harvesting.

You know it happens that when you are in a family, the discussion between a man and his wives is often helpful. They begin to understand what you are trying to do, so in this line of conservation work, really, my wives, Lizzie and Constance, help me a lot. And I am happy because they also have a sense that water shouldn't just be looked at and just left to run, to flow off.

Because of my water harvesting, today I am famous. Farmers from all over Zimbabwe, from all over southern Africa – even in Ethiopia – know about me and my water harvesting. Really, when I stop to think about it, farmers have known about my ideas for years, and many farmers even had the same ideas themselves. But the settler government we had before independence, they did not like us to use our ideas. The agriculture ministry forced us to do many things to protect our natural resources. Only now, it turns out that some of these were wrong for our environment. Even the experts now say mistakes were made. But if any of us African farmers tried to do something different, something for ourselves, we had a heavy price to pay. Fines. Jail.

Years later, after our own people took over the government, we began to find out that what the settlers forced us to do was not always good for our natural resources. Our soil and our water. The most dangerous thing turned out to be the thing we all had to do. All of us. Big farmers, small farmers, rich farmers, poor farmers. That thing was the contour ridge. We all had to dig them, or go to jail. Few of us had any money to pay fines for breaking the government's rules.

Some people call contour ridges diversion drains. That is because they divert the rainwater out of the fields. If you ask some experts today, they will tell you – contour ridges were supposed to prevent erosion, but really, they were the cause of erosion in the dry parts of our country, like here in Zvishavane, because people, even the agriculture ministry, did not understand how to use them properly. To work properly, the ridges must divert the water into a harvesting catchment, but this was not done. And there was another big problem. Because our soil is dry and we have little rains to fall, the natural elements of the soil dried out from being exposed to the air. Without the natural fertilizers, we had to use chemical fertilizers, and this hurt our soil even more.

At my place, I didn't like to see the rainwater being diverted from my crops. Only the big farmers had irrigation systems and us small farmers, all our farms were in the heavy drought areas. But I had no choice. I could not use my own methods, or the methods of our ancestors. Many of these ways are becoming forgotten because, under the settlers, we were no longer able to use them. The government thought it had a better way and that way they forced on us. I had to farm the way the government told me to, told everyone. So again, I started thinking. I asked myself, "Zephaniah! What can you do to harness the small bit of rain that falls on your plot? What can you do to have water for your crops and your family? What can you do to keep water in your soil during our droughts that sometimes last for years without a rain?" That is when I began my water plantation.

I plant water as I plant crops. So this farm is not just a grain plantation. It is really a water plantation. Planting water in my soil keeps it alive. When people come here to my farm, I demonstrate how to do it.

What I have done is to dig many large pits on my land, within my contour ridges. These I always called infiltration pits. But one day, one of my visitors began to call them Phiri Pits, to honour me. It is an easier name to say than infiltration pits, surely.

So these pits are now scattered all over my field. When it rains, the rain and the loose soil fall into these pits. They don't go washing all the way down into the sea where my crops can't use them. And when the water ceases to fall, the rain remains here. In my field. Nourishing my crops and also my soil.

These pits saved my life and my family, surely, and they have also helped me to become famous. But long before that, they also landed me in a lot of trouble with the government. But I stuck with my ideas when many others gave up, and that is why today I am known by farmers and agronomists and researchers across southern Africa.

The way these pits got me in trouble is this. You know I said that there are no rivers here at my place, no rivers like the Tigris and the Euphrates? So I asked myself, "Zephaniah," I said, "how can you harness this rain and keep it here, on your plot? If you don't, it will just disappear!" Well, along the edge of my place, there is an old streambed. There was no water there, but it was a waterway when the rains fell. And my contour ridges spilled into it. It was just a depression, but when water came from that end, it became a draining area. A common name for this type of area is *dambo*, but most people around here use the name *guvi*. So I made some pits off this catchment area, because when we had heavy thunderstorms, we had flash floods. And in these floods, all the rainwater would rush away just as quickly as it fell, carrying off the soil with it. But this being a wetland area, digging pits here was a restricted activity. And so I was called to court.

Well, again, because of my thinking, my luck was good and bad. It was good because at this place of the *dambo*, the *guvi*, the soil has a lot of clay, so I sank an infiltration pit here. Then I began to use it as a sort of a reservoir, to hold the water in place and prevent it from washing away unharvested. No other people used the land downstream, so I was robbing no one else of water. Really, I was beginning to save water for my neighbours to use. With each rain, I harvested more and more water. And I planted bananas and reeds in that area to protect the soil. You know I have a colleague who talks about 'accidental beneficiaries' of resource conservation. He means people and resources downstream from conservation projects. He says that from my place, things downstream look very green.

Now the settler government restricted us from farming *dambos*, which also went against our knowledge. You see, before the settlers arrived, *dambos* kept us alive during drought. We farmed them during those difficult times because *dambos* naturally held water. Anyone, from anywhere, could look at *dambos* and see that they had seepage even if they didn't have water. So, when the settlers came, they too saw the richness of *dambos* for farming, but they farmed them in a wrong way. They used heavy ploughs and chemicals. Even though we used only our hoes, and were not doing damage, the settlers restricted everyone in Rhodesia from farming *dambos*.

And this is how I got myself in trouble with the agriculture ministry and with the court. My water harvesting, on that *dambos* near my house, was restricted and so my activities were reported by a Land Development Officer. When the government found out, I was made to pay a fine. I was taken to court in 1967 and charged to pay ten pounds. After paying it, I was warned not to continue farming the wetland. But I persisted. I went on and added quite a lot of bananas within the wetland. I was not tilling the land or turning it into a ploughing system. I made some holes and planted these bananas.

After some time, I was again reported by the LDO. And this time I was asked to pay fifteen pounds. But I went again and kept my bananas growing, kikuyu grass growing, reeds growing within that area. Well, of course the government didn't like this, and I was called to court again for a third time. But that third time, ah, I got courage. I said to the magistrate, "I think it would be nice for you, magistrate, to visit and see on your own the crime I am blamed to have done." Then he – this man was really an understanding man – he agreed. We drove together. When we came to my land, I waited at my house while he toured my land. He found that the wetland I was accused for was well protected. I had planted grass, kikuyu grass. I had planted reeds. I had planted bananas, which really challenged soil erosion.

The magistrate, after seeing that what I was doing was so helpful to the soil, he said, "Mr LDO, what is wrong with this?" The LDO said, "There is nothing wrong." "Then why are you accusing him?" the magistrate asked the LDO. He said, "It is only the information that I got from the agricultural extension officers. Now I think that the information the extension gave me was wrong." Then the magistrate said to him, "I think it would be nicer if you didn't take any action against anyone without first seeing their project. It might prove other things." For him, seeing my farm was really a challenge because he had not visited the place, yet he was judging my crime in court. The magistrate was really very sorry that he had previously charged me for what he did not know was good work. From that day on, I was really given the go-ahead with the wetland management.

This farm had no good water source. I started digging wells to source water, but I only had chance seasonal water. My crops were good when water was there in those seasonal wells. But when these seasonal wells dried up, also my crops dried up. I again took up soil conservation. Firstly, I made some pits off the catchment area, off the seasonal streambed. Then later I put in a well, a small well, you see, that's where we get the water from. Then thirdly I made some reservoirs and ponds.

I was working on my own because most people were arrested for doing it in a wrong way. They used oxen to till the wetland areas whereas I planted some grasses to prevent soil erosion. Quite a lot of people were being arrested for misusing the wetland. This was asking for trouble, and as I explained, I was taken to court three times. When the magistrate finally visited, I think it was in 1969, it was after the good rains and my bananas, reeds, kikuyu and also my crops in the field were very nice. This gave a very good example.

I had made some sand traps, infiltration pits, dug big reservoirs on my catchment areas. Before I got arrested, the ministry of agriculture had already introduced contour ridges. Those were the only things they introduced. But these others, I just made on my own. The structures gave great help to my water source. I found that I had much more water for my crops.

The settler government made our lives very difficult, surely, but not all the Europeans who came to settle here in Africa were unkind. I had many good settler friends throughout my life. One of the most important was Mr Garfield Todd. He came here with his wife, Grace, in the 1930s. They came from New Zealand. Todd came here to lead Dadaya Mission, but he also became our prime minister from 1953 to 1958.

Dadaya Mission is near my home. How I got to know Mr Todd was through my father, Amon. The two of them were very good friends. My father came here from Malawi, only it was called Nyasaland in those days. My father worked with the Todds as a primary school teacher at the mission. Mrs Todd was one of the best teachers ever to come to Rhodesia. My father was also a preacher. He taught me everything about the Bible.

A friend of mine visited the Todds and they told her about my father. He was famous for his singing. We were all very poor in those days, too. Too poor to have hymn books for our church services at the mission. So, on Sundays, my father would lead the entire mission congregation in singing the hymns. He had a powerful voice. He would sing out the first line of the hymn, and all the congregation would sing the words. Just before the first line was finished, my father would sing out the next line, and then the next one, and that was how hymns were sung at Dadaya during my father's time.

Mr Todd really helped my family quite a lot; and not only us, he helped lots of people and is helping still. All the people around Dadaya and Zvishavane district really know Mr Todd as the light put up to bring all children to a better life. I remember especially when my father died, it was in 1950. The Todds had gone with their children back to New Zealand for a holiday and for a big meeting of mission leaders from all over.

When my father died, the mission cabled to inform them. Mr Todd was making a speech, talking about the work of my father, about the good work he was doing. And when the cable was delivered to him, he broke down in front of all the people and could not continue with his speech. That was how close, how good friends the two were to each other. It was hard, because just before they left, my father said to the Todds that, "We will never again see each other." The Todds said, "Of course we will be together again!" But my father knew somehow, had some feeling it would not happen that way. And this is what made it so hard for Mr Todd.

The government has always owned all the land that we farm. Well, in those days, men who were not born in Rhodesia could not expect their families to remain in their homes after they died. So when my father died, my mother and my brothers and my sisters and myself, we were told by the government that we could no longer stay in our home. We had to find some place else. My mother and all of us children were born in Rhodesia, but that did not matter to the government. We were all registered as foreigners because of our father. Well, we had no place else to go.

I was 23 years old then. My father had died, my elder brothers were away, and so I became the head of our family. I went by myself on my bicycle to see the Native Commissioner about the eviction, but he would hear nothing from me. It is fifteen miles from my home to the district capital, Zvishavane. In those days, it was called Shabani because the settlers had a hard time with Zvishavane.

But the commissioner would hear nothing from me. I left his office and went home. I talked with my mother some more about our problem. The commissioner said we should go to Malawi, to Nyasaland, my father's land. But that was not our place, we did not know it. And it was against our culture to return to the family of my mother's father. We had no place to go. So when I talked with my mother, it came out that my father, as a mission worker at Dadaya, had been negotiated for our piece of land by the mission through the commissioner's office. And that was how Mr Todd helped us. He convinced the commissioner of our right to stay, and this is where I have been ever since, thanks to Mr Todd.

My childhood was really very pleasant. I had some interesting things I enjoyed doing during my youth. We had some fruit trees, wild fruit trees, the indigenous fruit trees. I don't know in English what you call them, but I know them in our language. We had *mazumi*, we had *matowe*, we had *shuma*, we had *nhunguru* – quite a lot of species we enjoyed during our youth.

I also had some animals I loved most. These were my two dogs. They were called Bridge and Temba. Bridge had a white neck like a bridge so we called him Bridge.

The one thing I guess I loved most was swimming. We had a small river in our area called the Gwamatoro. This river Gwamatoro is named from being within the swamp area. So I really enjoyed swimming with other youngsters of my age. And also when I thought of leaving my dog behind I always found I was very, very lonely . . . yes, Temba.

So these were my very close activities that I liked, especially during spring we had quite a lot of fruit trees getting ripe, and I enjoyed collecting these for my mother. So when I could depart from my mother, this gave me a very good time.

Mazumi is in the family of oranges. *Matowe*, I don't know what I can say they are like, but they are just chewing gums. You keep on chewing, chewing, chewing, and the stuff in it is just very nice. You just swallow the stuff and the leftover, you throw it off. *Nhunguru*, in the family of peaches. Also *shuma* are in the field of peaches.

When I was a boy, I didn't know my family was poor. Neither did I know about money, about not being hungry when we had drought. My parents loved me and they took care of me and they taught me about farming. This was all I needed for a happy childhood.

But when I started to grow, to become a man, that is when I began to learn that things were very different for the black people who were always here than they were for the white people who came. I began to learn how my parents, how all Africans, really, suffered under the colonial régime. It was hard for us to get jobs, to go to school, to see a doctor, even to farm. Most of us were moved around, forced off our traditional lands and told by the settlers to, "Go there!"

But one of the worst hardships was that they used to take away our cattle. Destocking, they called it. This was not a punishment for a crime. This was a procedure of the agriculture ministry. They told us this would prevent erosion and save our soil. But before the settlers came we never had erosion.

I sometimes recall when all those things were so horrible, the destocking. If you had ten heads of cattle, the government would say you are to remain with two. When we were told that the animal is no longer wanted here, they cut – you know the tail of a cow has that hair? They cut short that hair. That's to show that this one shouldn't be grazed here. "Sell it, or you kill it," we were told. People would rather be arrested. They said sell, but we had no money to buy; we always traded. Our money was our cattle. And the government did not pay us. And they knew that our cattle were important to us for more than just food. We never slaughter our beasts for food unless it is for a very important ritual – more important than even a wedding. Only if one of our elders dies, then we might kill a beast, if we have one to spare, to honour that dead person.

No, to us, our cattle were our family wealth, our savings accounts. When we wanted to marry, our boys needed cattle to pay *lobola* or *roora* to their bride's family. That was our way of thanking a girl's parents for raising such a wonderful daughter to be a good wife and raise children to start our own families. Even today, some people still observe our ancient custom of *lobola*. But we use cattle for other things, too, nowadays. Sometimes we sell them to get money to pay our children's school fees or to buy some tools for our farms.

The settler government said that we had too many cattle in our small herds. But some of the settler farmers, they had thousands of heads. But they also had thousands of acres of land on their ranches and farms. Their cattle could spread out, the same way ours used to before the settlers came. But since the settlers came, our farms are very small, only a handful of acres. And they are not even ours. Our farms are government land allocations and the government, the old one and the new one, owns all the land. We still can farm only where they tell us.

I remember, ah, this was terrible, surely. You know those days were very bad, those days of destocking. It was really very bitter.

Whilst my father was a teacher on Dadaya Mission, we stayed home with our mother. And my father, when he left his teaching, he came home and did quite a lot of gardening. I was able to go to school at the mission, which really gave me the education I am with. I only went through my standard six. Seeing my father did not have enough money to send me to secondary schools, I just had to hang around and did quite a lot of activities at home – gardening. From my father's experiences, I also gained the way to handle the farm in growing crops and also looking after livestock.

You know, there was never very much water around here, a few big rivers far away, and no lakes. Because of this, local people never had the custom of eating fish. My mother, Magrate, she was Ndebele; she came from Hope Fountain. But my father, Amon, he came from Malawi; he was Chewa, and that is how I came to know the Chewa language of Nyanja. Anyway, in Malawi, there is lots of water and lots of big lakes. So the Chewa people were very fond of eating fish. My father loved it, but my mother! She could not even stand the idea of eating fish. But today it is different. The little kapenta fish that are now harvested in Lake Kariba are enjoyed by everybody in Zimbabwe, and we also enjoy the fish we catch in our community dams and fish ponds. And here, at my place, when people come, they see fish and eat fish that come from my own pond. Ah, I wish my father were here to enjoy my fish; he would be very pleased.

During my youth, my parents gave me quite a lot of duties, the reason being that I was the only man – a young boy, really – within the family. It was after the death of my brother and another brother was away, so all they expected from my older brothers was now on me. So I had quite a lot to attend to. I looked after cattle, donkeys, goats and also helped in farming, in spanning oxen. My mother used to bring me food as she knew I was so busy in the field. So this was really quite a lot for me. From ploughing, I would go up and do the milking. But all this brought much attention from my parents that I was really responsible for the family as such. In 1941, when my father told me about going to school at Dadaya, I thought, "Who could look after these animals?" Then I said to my father, "I don't think it would be nice of me to leave you because I see there is no one to look after our animals."

But about a year later, I went to Dadaya Mission with my father. My father was a teacher at Dadaya primary school. He introduced me to Mr Garfield Todd. Mr Todd was informed by my father of my activities at home and that I was most interested in looking after animals. "Fair enough," said Mr Todd, "well, the young man will look after the calves," so I found that I had no problem this time. When I was told I was going to look after the calves, I found that my interest was involved.

I looked after cows at Dadaya Mission during the morning hours when I attended the afternoon session, and did the milking before it was 12:00 p.m. When I went to school for my classes, I could report if the cows were not in on time, also inform whoever would look after them during the afternoon that I think they might have gone astray at such-and-such an area.

I went to Dadaya primary school for my standard six and that was as far as I could go because our family had no money. I grew up under Mr Garfield Todd's leadership. I can claim whatever I am able to do was taught me by Dadaya Mission. The English I am able to speak to anybody I come across is really a result of Dadaya Mission.

After all the trouble I had with the court in the 1960s, by 1973 my good work had at last brought friendship again. I was asked by the ministry of agriculture to join a group of farmers who went to Makoholi Institute, where a lot of erosion control activities were being conducted, for a course. I went for courses in poultry, gardening, fish and beekeeping, and fruit trees. When I came home, I put much time into these activities and improved. Beekeeping was very good. Vegetables growing, fish and fruit growing – well done!

So I came up with quite a lot of activities during that time whereby the commissioner wanted to see what good I was doing. The information leaked to his ears that Zephaniah was doing quite good in gardening and good in water harvesting. The commissioner wanted to let other farmers within the district come and see what I was doing. So my activities became very common to the communities of Zvishavane district.

So it was a chance that from my struggle with the wetlands I had now got a chance to show from my activities that I was not a person destroying the resources but a person restoring the resources. People were able to come and see fish – eat fish! – when they came to my house and this gave a very strong impression of my activities.

When farmers from other districts heard of my activities, of my successes with water harvesting, a number of them flocked to my place. They flocked to see what these infiltration pits were. They had never heard about it. So I took them to my home and showed them what they were like and brought out the measurements, four metres long, one metre wide, two metres deep.

I recommend these pits because when we talk of rains, of our rainy season, we may say November, December, January, but that whole season might come up with only one good rainfall in our area. For the rest of seasons then, I demonstrate to people that when you have contour ridges and have infiltration pits, the pits are going to hold the water within the catchment and it will seep into the soil. That will help crops more than letting the water run off. So this idea has made quite a lot of farmers very successful. This helps farmers really appreciate the seasons very much because they know that during that period, they are going to harvest quite a lot of water for their crops and also for their fruit trees and so forth, even the grazing. The cows are very strong because water shall really be found in the soil.

You know when I experienced that I could be a man of better life when I kept my resources properly and maintained my water harvesting, I knew I could be a man of better life than going for work somewhere. With water harvesting, I can support my family better than when I worked on the railways.

# Part Two Chimurenga

For a few years after I was at Makoholi Institute, things went along fine. But then, one day, the fourteenth of August, 1976, things changed. I want to say to you – I want to say God is there! – for on this date I saw something that I will never in life forget.

Freedom fighters had come to my home and asked me to keep their war arms until they could return to collect them. After some days, things did not go as they explained. On the fourteenth of August, 1976, I was arrested by the police and indeed got into a pot of fire. I was taken handcuffed and foot-cuffed. That means leg irons.

I was taken to Shabani police station where I was tortured. I had two of my shoulder bones broken, my hip joint. They hit me with the butt of a gun. I had not taken the war arms to fight. These weapons were left by freedom fighters under my care. But to reveal the presence of these war arms for me became a very hard thing. I knew if I revealed this information, the freedom fighters would have no cells to keep their prisoners; they would just shoot me. Then I dared say, well, if the government arrests me, I will serve my sentence and come out. But if I reveal this information, the freedom fighters won't excuse me, they will just come up and shoot me. Then I couldn't reveal the information.

The prime minister at that time, Ian Smith, he thought that the only people who cared about independence were a few people who had left for other countries to become trained as guerrilla fighters. Smith always called them terrorists, or terrs. But really, politics was something that had gone out to every district. Everyone was involved with *chimurenga*, with the liberation struggle.

I didn't know these boys. But when they came in with the weapons, and having been informed that freedom fighters are within the district, well, we were addressed by our politicians, our leaders, that, "Look here, you should look after these boys when they come." So with that, I had to look after these boys.

You know, this disappointed the government very much. You know the friendship I had mentioned, they thought I had really reformed from politics. But when they found that the freedom fighters had left their weapons with me, they just said, "This is a very bad man."

Some of the freedom fighters were arrested, and they were the people who revealed the information. From the torture, really, I couldn't accuse them. Because I saw from the time I was tortured, that the boys revealed because they had gotten really heavy tortures.

Now the hardship I had was that I was put in leg irons for almost five years. I was imprisoned here at my home and a flag was flown outside my house by the government to signal to the freedom fighters that I was a sell-out and that they should kill me.

But this is what happened. When the freedom fighters came back, they saw the flag, but instead they read in their minds that, "No! If this man was a sell-out, why would he be in leg irons?" You see the freedom fighters had long been informed by some *chimurenga* supporters. Mr Garfield Todd had especially gone into this. He had informed the fighters of what war activities were really like in the rural areas. If the freedom fighters were not aware, they would kill their own people because those people who were found being targeted by the government were just left to be killed by the fighters.

So this was my chance. When the freedom fighters came in, they had information of my problem and they didn't kill me. The freedom fighters knew that the Smith régime was doing tricks to make us fight each other.

But they wanted to take the leg irons. I said, "No, I am now a cripple, I can't walk," because they wanted to take me out of the country for treatment. I said, "No, I can't walk – please – save my life. Leave the legs irons and leave the flag." But they wanted the flag and I said, "No, once you take it, you are targeting me to get more problems because I will be asked again why I have not reported the flag." But they broke the pole that held the flag and left the flag with me.

Then the government security forces came and took me, and I got quite a lot of torture again. They hit me. Today my hip is not up to date because of those pains. I was taken to prison.

I stayed for three months at Gweru prison without being charged. I was in isolation for three months, always in a cell with just a small window. And then they brought me home when they saw that the torture was showing that I could die any time. I was tortured and I think my blood got hurt. I had quite a lot of bleeding and no treatment was given. Both of my eardrums bled very much and my testes were hurt.

The torture is still in my body even today after twenty years. I am sorry to say the problem of me losing my job brought no education to my children. Brought my wife, Lizzie, a big, big problem. This was a very hard time for my family. I still have a big problem that I never forget because it is something that is really affecting me, the problem of education. During the struggle, I spent most of my time under restrictions, detentions, imprisoned just because I had some feeling to keep ourselves alive. Then this meant that my family really did not achieve education as such. So the question of education is something that no one can deny, but this problem really affected my family very much.

Divided people. Now the people were divided. When the government saw that I had all these weapons, they came to address the community and they told the community that, "Look here, Zephaniah is a very bad man." With those weapons, they said, "Here, he brought these to kill you!" Then they demonstrated the AK-47 on a tree and the tree was cut down. So they said, "That's the idea Zephaniah had. Do you support such a man?" Then the people weren't having me. Divided. The government said that people should not communicate with me any more. Nobody wanted to visit me.

The time became very tough. When freedom fighters came in it was a problem to cook for them. This meant quite a lot of problems for the community. Well, some people were beaten, some people were shot. Some peoples' homes were set on fire. The cattle, everything, were killed by the régime. So all this became a really bad time. When the régime heard of the presence of freedom fighters within that area, people were brought problems, yes.

And when it was found out that, "Yes, we did cook for them here," all those people were killed by the régime. But these freedom fighters, seeing they had nowhere to get food, they depended on the communities. We are the people who cooked food for them. But the moment the government heard that some food cooking was done at such-and-such a place, those people suffered a lot.

When the freedom fighters came around and saw that there was no food, people would be beaten. They trained people to go and get cattle from farmers; people were taught to go out and get cattle. They taught people – told people – "Go into that farm and get some beasts to kill!" All this was done by the people, not willingly, but, well, to save themselves. To save their lives. They had to.

The régime did not want people to support the fighters. If found, helping or cooking for them, people died for that. Some got their livestock killed and homes burnt down. People could not have a safe side because freedom fighters also killed them. Killed those who were found helping the régime.

Times became very hard in Zimbabwe.

When I was at Gweru prison, a strange thing happened. One day, my cell was open. They found my cell open. I found my cell was open, too. But I had leg irons. There was a chain from my leg irons to a ring in the floor. The floor chain came straight up to my bed where I slept. So when the guards came, they found the door was open. They asked me, "How come? Who opened this door?" I said, "Look, it is surprising because the cells are not locked from within, they are locked outside. I don't see the reason why you should ask me this question because these cells are closed from the outside. You should ask the people outside."

Then after three days, this thing happened again. The door was open. Then they said to me, "What church do you belong to?" I said, "I belong to the Churches of Christ." And they asked me, "Do Churches of Christ believe in spirits?" I said, "I do not understand your question because whoever worships worships the Spirit, so I don't know what you mean." Then they said, "Are you possessed in your church?" I said, "No." Then they said, "How come the door has now been found open twice? It must be terrorists," they said. "They want to come in and snatch you off." Then they said, "You will stay here in leg irons and handcuffs." I said, "Well, it's all up to you." After two more days, they said, "No, you must go back to Zvishavane."

I didn't like to come back to Zvishavane. Because of all the torture, I didn't like to let my family see me any more. I felt it was better just to die isolated. But they just brought me out. When I came out, I found that my eldest son had a violent attack with the commissioner. He quarrelled with the commissioner that, "I want my father!" But they sent word to my family that, "You will never see Zephaniah any more – he is dead!"

But the very evening of that morning, they drove me home. It was a shocking thing. When I arrived there, I didn't know my brother had died. So when I arrived, the whole family was mourning.

This was when the Native Commissioner and the Special Branch of the police made a party for me. They celebrated – well, you know when one dies in our tradition, after twelve months a special party is prepared to celebrate that so-and-so has died for such a year, so we want his spirit to come home. This is called a *bira*. The spirit is called in to come back home. So this was done to me, as though I were dead.

I was told personally by the commissioner and the Special Branch that, "Look here, we want you to celebrate your *bira* before you die because you will soon be dead! So we want you to celebrate your death – to celebrate that you are dead! So that you will not worry other people to brew beer for your celebration." They brought drinks and biscuits for people to have at the party.

Just a few people around my place attended. They were collected by the commissioner's messengers. A small group came in. People were so frightened. You know I was found in possession of the war arms and the soldiers demonstrated how the gun was used. So people feared to come. They also feared that if they didn't go, there would also be a problem.

That day they put me in leg irons – ah, Zephaniah. And they tied me. Even today, this part of me, my bone here, is very painful. They put on the iron and the handcuffs, and they were very tight. My arms were pulled back behind me.

I think the most important damning issue they had was how did I happen to have the war arms. They really wanted me to reveal how, to reveal where, I had got these war arms. So that day, they put me in leg irons, a party was fixed at my house.

I wore those leg irons until independence – four years and five months.

One day I was telling the story of this *bira* to one of my visitors. She asked me, "Why would you die from having a soft drink?" How could a soft drink kill me? It was poisoned! By that time, the government security forces were secretly using poison on us. A good friend of mine was not so lucky. He was given a soft drink to drink and he died.

This was another one of the government's tricks. You remember I said I always had some settler friends? Well, one friend, he saved my life, surely. Mister? – what's his name? – I've just forgotten his name! He was a white man, he was really a friend of mine. His name, it starts with – ah, I've just forgotten!

But this man, he got information that the Central Intelligence Organization had decided to give me a poisoned drink so that I could die, because after the torture, my days were really short now. I do not drink those beverages. I only take tea and milk. But at this party, they brought a bottled drink to me. They opened the bottle. They said to me, "Have a drink." I said, "No." Then one policeman whispered to the other, he said, "He thinks it's poison." Then he said to me, "Do you think we put poison in it?" He didn't drink that one he had given me. He took another one, he opened it, he drank that one. Then he said, "Take this one and drink." I said, "No, thank you." But this friend of mine ran at night, and he warned me that I should never take anything.

I was given a poisoned drink so that I should die! That's how I really came out to survive. It was due to my friend, surely. Otherwise, if I were a person who drinks bottled drinks, I would already be gone.

God's guidance is really within my family, because, you know during the hard days, I always saw that somebody guided me, made me safe from getting hurt.

Part Three A Water Harvester's Gospel

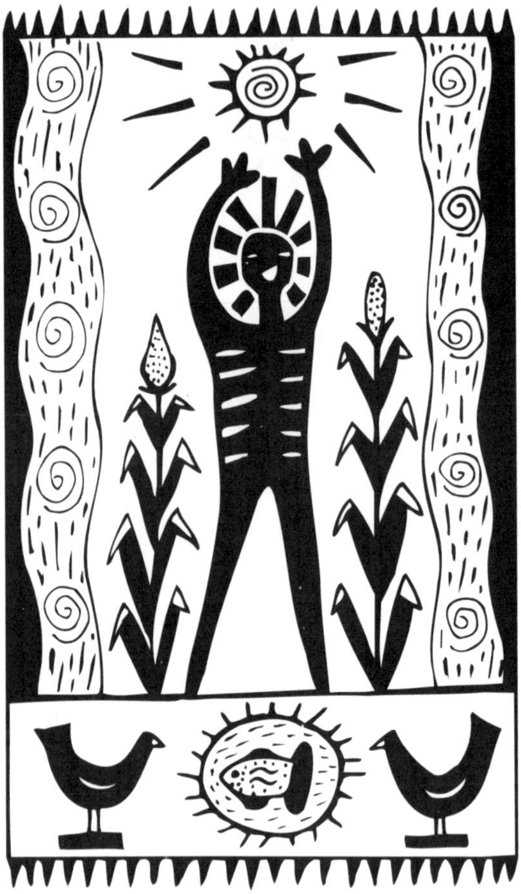

I call water life.

My trees around here seem to talk to me – "Thank you, Phiri, for harvesting all this water around here." The frogs all sing very lovely songs because when the reservoirs have water, they all sing to enjoy.

As I mentioned, this land is not mine. It is a government allocation. "Zephaniah! Come here! You are allocated this piece of land!" So if I monopolize all the water, I would be selfish. If there was another farmer, he or she would not achieve that chance. So I practice something that is really going to be accepted by everybody. No one should complain against me about harvesting for my own. You know, I even have to limit myself because I have my fish in here. If I take too much water, I will kill my fish.

I have a small orchard around which really gives me quite good fruits. My trees have no ditches for watering. To water them, I just use nature's reservoirs, the seepage. I look at the trees within the country. These trees identify that they get water in some way. And that some way I feel is within the soil. So, I say, my trees get water also from this reservoir naturally, not by me carrying water to them. Trees suck water naturally into the soil. I have mangoes, I have peaches, oranges, granadillas, lemons, guavas and paw paws. To the community, I give them seedlings to plant. It's the mango that has succeeded. So otherwise, there is no one who even attempts to grow any fruit trees at their homes.

It was me who started very nice drinking water and also asked people within our community to come and drink from my well. I also started toilets in our community. I also started to inform farmers in our community to utilize sound farming methods and informed them not to use fertilizers but to use manure. I introduced infiltration pits, which in my community every farmer is using. I also introduced intercropping in the fields and mulching, and explained why it should be done.

I try to prove the idea of self-reliance to local people. If you harvest water, you will run short of buying a car, but to help yourself, this you can do. And this is what I have done to help myself with life.

I started this water harvesting practice and really saw it was very helpful to me so I couldn't stop it, and even today I am still having some ideas to practice on this water harvesting. Really, many farmers do not put it across to themselves that they can harness and save water into their soils. That's something that people should really try and it's a success if you do it. I started harvesting water, the run-off from my granite, the granite *ruware* just near my home. Then when I looked at the water that I harnessed, I always watched below. The area remained very wet for quite some time.

The very serious thing I have noticed is that farmers enjoy farming, but they don't know the dangers that occur each rainfall. By danger I mean that the run-off water carries away quite a lot of our fertile soil, then the poorer ourselves become. This resource is our pride. Because with this resource, we have all that we want.

Here at home, we are able to show the community why we need a garden at home. We demonstrate the use of the dry land and the wetland because the people usually think gardens are at the wetland and not within the dry land. So here, with the water we have harvested from the rock, we are now able to use it for our summer garden. So the community is really able to understand; this is a place of demonstrations.

I built a small tank so that when I talk to farmers they can really see what I'm talking about. When I say, "Let's harvest rainwater from our roofs," I can show this water when I talk to them.

Here at my farm I have made a very good chance with other farmers. My experiments show me that when rains fall, I shall harness all the water over this way, through those sand traps. Three or four days after a rain, when the land is getting dry, I call the farmers, "Come and see! I have water here!" so that the farmers can really be sensitive to water harvesting. When you come here, you'll see the water moving into this pit. Then I often ask farmers, "Why is it that the water is seeping in there? Why is it running that way?" When I go to look at the soil I don't see water because the seepage of the water is within the soil. But when I damage here, I see that God is against opening up this ground. The purpose of the seepage is that God wants this area filled up. So then, when I am also cut, you will see my blood oozing out. But the purpose of this blood oozing out is to clot. Once the clot closes, it will be healed. So the seepage shows that nature says, "Recovered!"

So that's why I say nature – God – created man just the same as the Earth.

On my farm, I have a place I call the divorce point. When I look at the happenings that take place there, I see that I am divorcing married people – the water and the soil. They travel together. When the rains fall, these two run together. So here I made a divorce point. With a sand trap, the water is slowed, so that the soil it carries falls from the water to the ground.

You know when you drive through a town and you come across some big humps in the road? Those humps say to the driver, "Reduce your speed." So that is what I am saying to my harvested run-off water: "Reduce your speed!" That way the water is divorced from the soil. It has been travelling, but the soil remains. When the soil remains, I look at that leftover soil and see that it is unknowing. I ask it, "Where were you going my soil?" No answer. I ask again. It says, "Well, I don't know. Somebody has carried me." That somebody is water. Then to this sand I say, "You are not allowed to travel with water. I will take you and make use of you." Then I take it and put it just near the bank of my reservoir.

Water can talk.

I make this a place where water can speak to me. Some of the water is frozen among the stones of my containment pit. When I come here and look, I find that the water is saying, "What does he want me to do now?" So I say, "Please, go over there, take that way." The water will flow into that pit. So water obeys my order. This water here I use for my vegetable gardens.

I have made a place for washing plates. It's most common for whoever is washing just to take the water and throw it away. Then I say, "No! At the Phiri place, water is blood." This water we use for dishes should also be taught, told where to go. Just after washing, please, let the water flow into this pit. The water comes here, you can hear the hollow sound under the ground here when I stomp on it. This pit here shows that even the dirty water is very useful.

Here I have an underground tank. It is just a natural tank. I dug a large pit. Then I filled this pit with stones. I put in some big stones so they can leave some gaps. I didn't put back the soil I dug from this pit. I just stuffed the pit with stones. I put a big one here, another big one there, and then another comes on top. This pit is filled with staggered stones, and then I covered the stones in the pit with a sheet of plastic, and then with earth, and I made the ground smooth again so I can walk over it. People do not know this pit is here until I tell them.

The spaces between the stones allow a lot of water to fill in between them. Then in the pit I put a breather pipe, because when I siphon water in, there is air. Without a breather, the water won't go in fast. So to harness more water, the pit has a breather so that when the water comes in, the air goes out. A pipe channels the water from my catchment into this pit so that it will go underground and stay there, nourishing the soil. Being underground, is it protected. It cannot evaporate.

So this pit is really a tank. I call it a tank for the poor man. Because I don't enjoy when a farmer says, "Well, I can't build up a tank because I have no resources." Then I say, "No! Just build! Make it from these common presents! You were just given this Earth by God. Just open up, then allow the water to go in."

This pit is not lined with any cement. It's just earth. So the water will be able to seep into the soil, not letting the water flow over. This is a very big challenge. You can see my trees here. There are no watering ditches in the ground around their trunks. I have seen quite a lot of farmers digging ditches around their trees. "You pour water there? Ah!"

I sometimes think no, there must be something wrong with my fellow friends because I don't imagine the roots of this tree being around here. Some of them are far away. So if I plant water into my soil, my trees' roots are going to travel to fetch water.

So that's the idea I wish to throw to my fellow farmers. That they should really make some efforts to plant water.

You know that streambed I mentioned? Well, today that seasonal stream is a pond. It is quite large, about a quarter of an acre, and the water in it was harnessed over the years by my water harvesting ideas. A pond is a rare thing in this part of our country and many people have come to my farm to see it for themselves. It never goes dry. During drought, people walk as far as five miles to get water here.

And I can now also use its water to irrigate my crops. With my wives and my children, over the years we have dug a web of ditches into our plot. The soil around my grain field has quite a lot of clay, so this clay preserves the ditches.

But I have also done some engineering. I observed how the water flows naturally down the slope from my pond. So at an intersection in my watering ditches, I installed a small piece of pipe under the clay. This pipe connects two fields. I took some old rags and stuffed them into a few plastic bags and then I tied this into a ball. When I want my crops in the lower field to have a drink, I pull this ball, this stopper, out of the pipe and the water flows down, slowly into the ditch. Because the ditch is natural clay, it can breathe the water into my soil to nourish my grain with the pond water. When the crop is no longer thirsty, I replace the stopper. The natural suction of the water keeps it in place and no water runs away to be lost.

Another thing I like about my pond is that when farmers come to visit, they can see fish and eat fish caught here at my own home. And of course my wives, they like it very much. They are some of the few women in the district who have water at their home. Most women in our region must walk many miles each day to fetch water from a well and tote it home in a bucket.

I have already mentioned my infiltration pits, or Phiri Pits, as my friend calls them. These are not just holes I have dug into my fields. I call them structures. Farmers with soil that is hard and dry, like most of us around here, can use these to allow rainwater to infiltrate their fields, to build up their water plantations. These structures cost nothing to build. Anyone who can dig a hole can build one.

I know these pits will keep my rain very successfully. My crops won't dry up. With contour ridges alone, the rainwater flows out of the field and disappears if it is not properly harvested. But with the pits I have dug in my contour ridges, the water that remains in my pits remains my water for my crops. It stays in the ground and it does not evaporate.

Farmers come here and I demonstrate. I show them how I do it, that's why they flock here. And not just farmers. Researchers and students and government officials from all over the world come here to see my water harvesting and also to hear some ideas I have discovered on water plantations.

I plant water as I plant crops. The water harvesting I do, to me it's really a plantation of water into the soil that proves to my soil a success. Yah. Planting water into my soil, using compost in my soil, using leguminous crops in my soil – that makes my soil remain alive.

Erosion is not a simple thing. It is our soil washed away. We need that soil to remain in Zimbabwe. In many places, it is being washed away, right out. I think it is an emergency.

From my own experiment here, with this idea I am doing, I never run out of water, and in drought, I invite my neighbors to come here for water. So I think with this idea, farmers, if we all really could adopt it, it would help us a lot. Because even though we say our region has little rain to fall, if we only happen to harness that little water and make it a point that this water is planted into the soil, surely we will go through the problem of droughts. Yah. Zvishavane Water Project is trying to introduce soil and water conservation to the whole country. But the problem now is that farmers just sit idle and think the sky is going to bring water in. That's the problem we have.

When I look at southern Zimbabwe, the problems we have are not very new. When I look at them, especially when I look at the rains, the rains, they are otherwise the same rains. Only now, the way we are supposed to help ourselves becomes clear. I am giving awareness to the community that they should not let water run off but that they should try by all means to harness the water within their premises, within their contour ridges, their catchment areas. That is, farmers should make up some structures into which this water could be led so that it will seep into the soil. And by so doing, we are helping our water table. Our water table will always keep quite close to us.

After the war, I was invited to work with a group who had come here from America to help us with our water resources. I was introduced to this group and they hired me as a supervisor. They saw that I could sink wells so they wanted me to join them. And I joined them. I gained a lot in learning about the construction of dams. I did quite a lot of activities. I could supervise well sinking. I could supervise gardening, also damming. But this group didn't work in our area for long. And they left quite uncompleted projects. This didn't settle well in me and when they left, the people all around the district knew me very well and they said, "Phiri, look at that well; look at that dam!" So this made me feel I should try and hunt for money somehow, finish up these projects.

I want to thank those who gave me the go-ahead with the Zvishavane Water Project. It was a chance that I had a young friend who never hid the information. He told me of the World Development Movement in Britain. This was in 1987. Then I wrote a letter to the World Development Movement saying that I had made some researches of my own, that if one is able to keep his soil, is able to harness water, then this person has life. They wrote back and asked me to come to Britain. This was a challenge, really. When I read the letter, I found that, "How could I reach England?" I had no money. I wrote back and told them I had no money, but I was very equipped with the information to give them.

They managed to make me a ticket and everything for me to use being in England. And I was called to come up in June, 1988. So I took off to England. When I got to England, to me it was a surprise. "What am I going to give these people?" But from Zimbabwe I had brought along slides taken by my friend. When I got there, I found out that England had water; it was always green. So I found the slides were really going to work well because I come from a dry area where there are erosions. Where I looked in England there are no erosions. So these slides gave a clear understanding that I would really have my problems properly attended.

The first thing I did, I was taken to a meeting with people who really wanted to know who I was, why I had come in, what I was carrying for them. So I displayed all this. I just asked for a projector then I put in the slides. The slides spoke aloud, and they were very well heard. So I did all this – eighteen meetings with these slides.

Lastly I was taken to Oxford where I came up with a success. Oxfam (UK) and EEC Microprojects took me with green hands and gave me support, and another donor, NOVIB, has also joined us recently to support our work.

So today I have a project called Zvishavane Water Project, which generated from my own house project. I am the founder of this project, yet I have felt through the initiatives of Oxfam that I shouldn't remain only in Zvishavane and Chivi districts but that I should go wider than that as Oxfam needs my help to be taken into other countries.

I can carry the message anywhere.

I got the chance. I had already built these structures of mine. When I came in with this money, I started to look for other people who would like to join me. That's when I started to have an office. So it is from my trials with water, I got convinced that this can help anyone survive.

What I enjoy from it is that I was arrested sometimes for using part of the wetland. The government, the colonial régime, didn't like anyone to use the wetland. But when I used it, and brought up quite good harvests, then the régime came in and asked me, "Zephaniah, would you allow us to bring visitors to come and see what you are doing?" You see, then, to me, I saw that, yah, this was the chance. That the régime is now able to understand that Zephaniah has a crop that he works somehow.

Then I started to receive friends from outside quite some time back. But seeing that I am not that educated, I'm not recognized, because they say, "Ah, what does he know?" But I am happy because people come in and learn from the uneducated man. My joy is that I got the idea from the Book of God, and that the rivers – that the secret people never come across, never understand, is why the river, and why the Garden of Eden was put there as a source of water. People do not think about it, but I thought of it and practiced what it is. And today I have fish at my home. Not at the river, but at my home pond – I have fish! My children know how to hook just from my own fish pond. So to me it's really a joy because I was given the energy to do it by God.

I wish the people could see the forecast I have in me. As the Bible says, you are the salt. When Jesus sent his disciples, he said, "You are the salt to salt the food for the people." I'm saying, let Zvishavane Water Project be the salt to show the community a better way.

I went to Ethiopia for the Globalization of Agriculture and Growth of Food Insecurity Conference where I was asked to inform the conference of my home project. I took the chance of the conference. I made everyone listen to my speech. I said, "To me this project was a call from God, for if I was not called to take up this project, as I have not been to agricultural school, this could not have come in this way." I told them what my project looks like and how much it has helped many farmers from all over the world to see my water harvesting ideas. You can see I had a very nice time in Ethiopia. I have come up with a chain of visitors' names from other countries who want to come home to see for themselves.

I have a colleague who reminds me that a prophet is useless in his own context. I am not a prophet, but I do like to set an example that people can follow, and this I like to share with my fellow field workers. In our monthly staff meetings, I sometimes lecture to our group of workers. I say I think it is nice to understand that the projects we carry out belong to the communities.

Our workers should always understand that the projects are not for the leaders or the officials. The projects are for the communities. If I die today, let's say I die – Zvishavane Water Project can't just stop because Zephaniah is not here. That's why we really must inform the communities about awareness.

The people should voice. I have a feeling that we should really explain to the community to have power. The communities should have power over their own projects, all their projects. If I say, "No! I don't want you to make a dam," should the whole community stand idle? Do you feel that I have a right to say to the community, "Don't do it?"

So it is up to us to educate the communities on empowerment. We cannot fight off that truth. We are servants to the people and the people are our masters. Let the people voice what they really want.

The poorest person, that's the poorest person we want to attend to, not the rich ones.

Communities look to us, to Zvishavane Water Project, for help, and this we are committed to giving. But our water harvesting work cannot be done without conservation work. For example, community gardens.

Any land that is to be used for tillage needs contour ridges. And the contour ridges need infiltration pits. These work together to harvest water. The run-off water, most of the water that is washed away through the contour ridges is collected by the pits, and that water seeps into the soil. And that fertile run-off soil is not carried away.

And dams. If a community needs our assistance on damming, the first priority is to see that sand traps are made. Sand traps are walls made of broken rocks that we clear from our fields. The rocks or stones are piled on top of one another to make a wall. The traps can go across a gully to stop the water. Or they can be put wherever water and soil are washing away, like on a rock slope, a *ruware*. The sand trap slows down the water, and the soil it is carrying settles to the ground. They have not been allowed to wash out. In some places, many sand traps can be built to form a terracing system.

There is another very important thing about dams. Dams cannot be started in the rainy season. From October, November, December, January, core trenches for dams should never be dug.

At ZWP, we are discouraging maize cultivation. Okay, we know maize is easy to grow, easy to harvest, easy to cook, easy to eat. But the problem is that it is unable to stand against drought so we encourage farmers really to look into our traditional crops such as our millets and sorghum.

We are also discouraging chemicals. God himself created Earth and on this Earth he never put artificial fertilizers as we see now are being made. These artificial fertilizers damage our soil. I wish farmers would really understand the importance of using nitrogen-fixing plants in their fields. These would serve the soil for quite some time. I don't know why the agriculture ministry encourages farmers to use artificial fertilizers, because the moment the soil is fertilized twice, or three times, the soil is no longer strong. Instead people should be encouraged to use composts.

I told my fellow ZWP field workers long back, "Be exemplary! Be exemplary as a field worker!" People look at you differently if you say that you should put a tie on but you don't have a tie on yourself. They will ask, "Why does he ask me to put a tie on when he does not put a tie on?" Then that's how it is. If you always preach to people to have infiltration pits but you don't have them at your place, people ask themselves, "Why does he want me to have these pits when he doesn't have them?"

Also in our staff meetings I sometimes ask our field workers to question: "Why are your projects failing?" I say to them, "Do a really strong finding, make quite a lot of research over what you are doing. Is it your approach? Is it that *you* are failing?" If so, ask fellow field workers to come and help you.

We are really needed to spread the word to look after the soil. Because soil is the base of all our activities we do. If we just let the soil wash away, that means all our resources. It is the duty of Zvishavane Water Project to see to all these problems.

There is another thing I do if any of our groups are having problems with their projects. There are some projects that seem to fail. I go to meet the community and discuss their problems, why, what it is that makes them fail. I sometimes take them to my home project to let them see. Then from my home project, they go back with the idea of doing what they have seen.

I'm trying to strengthen the communities. I often like to network, bring groups from other areas to come and see what the failing groups are doing. Sometimes I find that certain groups do a very good job. Then I ask the groups that do a good job to come and assist the failing groups so that they grow relationship. They should know they are a family of the Zvishavane Water Project. So that influence really helps groups to grow their relationships.

I'm a free man who wants to help the development activities. We are a development organization and it appears the Zvishavane Water Project is becoming very important. But looking at my own expectations, I wish that my feelings could go inside every field worker. So that the field worker should not feel that he or she is working for money.

The field worker should feel that he or she is really spreading the information, the truthful information to the community. They should also have the feeling that their duty is life to a generation. You have to train people, draw them, call them, ask their husbands to come and see. Just ask them to come up and see, and then from there they shall have learned a lot. We need to show our determination towards the successes.

Water harvesting mostly involves women because women are the people who carry the water from far away to home. Then when you talk about water harvesting, women really contribute quite a lot because they know the problem is almost solved when water is available to them.

At present I find that people have a light. You know, in the past, in the colonial régime, people were never informed about their own problems. They were taken only as forced labour, "You have to work here!" But they didn't know why. So this time, this education is bringing people to a light that it is they who have the problems, that they should really work to achieve successes from their problems. So this time you find that, "Yes, Phiri is here to help us. People, let's involve ourselves." But sometimes groups or communities may have a strong problem with food. Then that's where you will find the delays on the project are caused because people go out and try and scratch for food; to get money to have food. This is one of the problems we have come across. But at least really everyone now understands that he has to do his duty to achieve good.

Harvesting water is really favourable to farmers when crops do well. But if water harvesting becomes waterlogging, it becomes a problem to crops, and farmers sometimes think: "No, the water is too much." They don't want it. They need average water. So during the rainy years, those farmers who took my ideas of making some infiltration pits, when the rains come quite heavy, the infiltration pits hold quite a lot of water. So some farmers thought, "No, it is no good to have them in a year of good rainfall." At my place, it is in a dry zone, so if I get waterlog in the rainy season, I will still have the seepage from my pits to get good harvests in spring and winter.

But during the bad years, farmers really say it's very useful. I have quite a lot of farmers who support the idea of infiltration pits. So I'm sure the infiltration pits are very much supported by farmers. These people believe they can help themselves by harnessing water. They invite better harvests.

God has given to me some blessings. My children have seen my vision. I told them in my dream I had water in my field almost covering my crops. And when I got up, I told my children. And this dream has come as I saw it. This year, my crops are bad from waterlogging. But the water harvest is wonderful. In past years, we never had such rain in history. When I think about water, I always find it a precious topic. I enjoy harnessing water. Really, you know, when the rains fall, and I see water running, I am running! Sometimes you will find me being very wet.

Zvishavane Water Project works with communities in Zvishavane and Chivi Districts. We help communities to harvest soil and water in a number of ways. Phiri Pits are one way, but we also help organize community gardens, fish ponds, harvesting tanks to collect rainwater that runs off school roofs and house roofs. In our area, we have no other water source but natural sources, so we must harvest it ourselves. We have no irrigation and we have no taps in our homes. Also, we have no electricity, no telephones and no roads. But with our water harvesting, we can get enough rain collected to keep our crops and our animals alive, and to get some fish for ourselves.

Dams are another thing that is very active in the communities. Some of my colleagues are critical of dams. They say that instead of making a water plantation, like with Phiri Pits, the dams allow too much rainwater to evaporate. Well, this is true. The dams are built within river beds, so the rain harnessed by dams stays within the stream line. It does not filter out very far into the surrounding soil as much as with a water plantation. But without these dams, this rainwater would flow away without being used by our farmers, our animals or by our soil. It would just disappear. Wasted.

At our ZWP projects, we do water harvesting together with our conser work. So if we help communities to build dams, we also help communities with conservation, to understand the problems of erosion. To work to stop it. To stop our water and our soil from washing out.

Our dams are made of earth and sometimes of cement. Zvishavane Water Project has now helped eighteen communities to build dams. So far, none have burst and there has been little problem with siltation. We use our technology very carefully to avoid these dangers. What we do is help communities to build sand traps above their catchments, and this gives a strong help to keeping sand out of the dams. And we do not just go in and start to work on a dam. We work together with the agriculture ministry and the Natural Resources Board to get the best site and to get it properly located. A dam, a garden, anything that is to be constructed on the soil involves the agriculture ministry.

Our dams are not huge structures like Kariba Dam. Our dams are all built by hand, by our hands. We lend tools to the communities and supply some materials such as cement. It usually takes about nine months to build a dam. Men and women from each community divide their time and contribute their tools, their labour and their animals so that it really is a community project. When the dam is finished, then we all just sit back and wait for a good rainy season so the dam reservoir can fill. Our dams usually retain water for the communities without ever drying up. Then there is water for washing and for our animals. Also, for fishing.

I enjoy seeing farmers making a trial experiment after visiting me. Then they invite me, "Phiri, come around." When I go there, I find that some of them have done quite a lot, better things than I have done, and I say, "Okay, what's your discovery?" Then they tell me that, "Yes, we have been wasting time letting all the water just run off. Now we have gained the chance of harnessing water." Then I say, "Okay! That's the success I enjoy!"

I have been visited at my farm by people from many districts: people from Berengwa, people from Bikita, people from Chilimanzi, people from Mutoko, people from Sanyati. And I have also had visitors from many different countries: from Namibia, from Swaziland, from Botswana, from the UK, from America, from the Netherlands, from Mozambique, from Malawi, from India, from Zambia, from Kenya, from Somalia. Well, all these people have really witnessed my water harvesting as being a very successful issue.

I once had a woman from America visit my place and I later took her to visit one of our ox-drawn, dam-scooping projects. Afterwards, the people there told me that in their lifetime, no white woman had ever visited their village. They said it was a blessing, and well, maybe it was. That dam project got finished and the very next rainy season was a good one, so the dam spilled in no time. The people were very happy. I later wrote to this woman to tell her the good news of the dam. She was shocked to learn that after a hundred years of settlers, only African women still knew that place.

So you see, when people come in, we keep the relationship, and sometimes they invite me to their places. I have been to Namibia and Zambia. And sometimes our agriculture ministry holds very big workshops in Harare or Masvingo where I am invited to address my activities to farmers.

I would like to inform road constructors about when they construct roads. We need roads, but I wish they would do some research firstly, see if they can make up a draining point, make up a reservoir so that the rainwater can be led into the reservoir, led to sink into the soil. When I look at the machinery that scoops out the roads, the people make heaps and heaps of gravel so I think there is a very reasonable chance of making us some reservoirs where they work on the roads. I think that this should really be brought to awareness. If we have constructors, let them also help us build reservoirs.

In December of 1998, I retired as the director of the Zvishavane Water Project. In February of the following year, a big retirement party was held for me. The district administrator even made a very nice speech about me and the work I have done with the communities. I retired because it was time to allow new people to take over this work. And also because I have many experiments that I am working on at home.

Shortly after that, I received word that my experiments would get the chance. I had applied to ASHOKA for some new funds to conduct my researches. And my application was heard from among a very large number of people all trying for support. So my work goes on at home, but I still continue to spend quite some time in guiding the Zvishavane Water Project into the new generation.

Zvishavane Water Project is an indigenous organization that I founded. So when I came to think of retiring, I thought it would be nice for me to give the Project a piece of land, about 0.7 acres, so that the Project could demonstrate to farmers some ways to bring communities to a strong self-reliance. So I am happy. After the offer that I made, Zvishavane Water Project has taken a step. They have started vegetables - tomatoes, onions, pumpkins, and so on - whatever crops suit the communities. This is really the heart of my desire. I have also visited quite successful projects especially in Zvishavane district and in Murehwa. We have quite a number who have started fish ponds. Zvishavane Water Project has really grown up to be a friend to most farmers. Farmers from all over - bus loads of people from all over the world - have come to see what it is like.

I have a cancer in me which I think could only be cured when Zimbabwe turns the poverty into happiness. Then I am sure I shall have achieved the cure of my cancer.

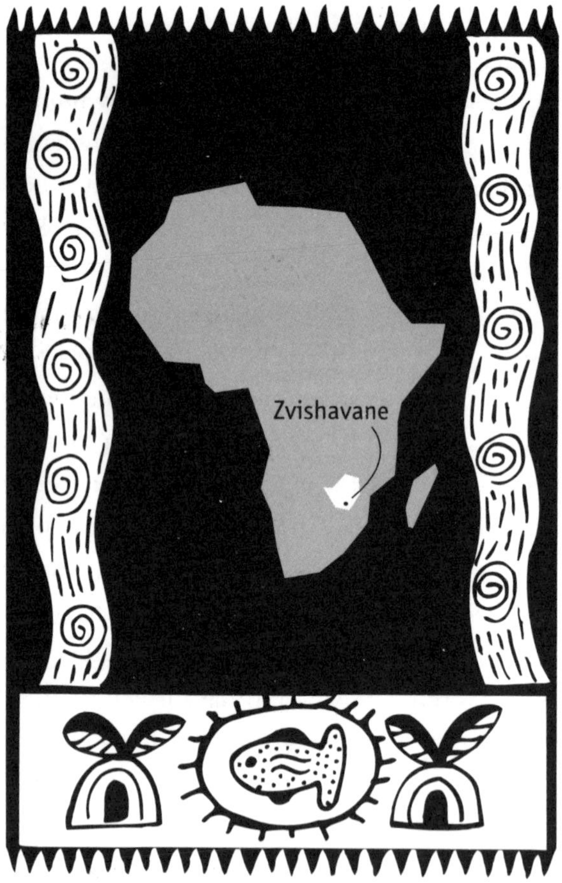
Zvishavane

Illustrations Mr Phiri's Water Harvesting

These illustrations serve only to approximate the dynamics of Mr Phiri's fields and aim to show a general working of his water harvesting. Mr Phiri's intimate knowledge and understanding of his land has resulted in a complex network of terraces, sand traps, infiltration pits, ponds, reservoirs, tanks and contour ridges. These drawings greatly simplify these devices and their connectivities.

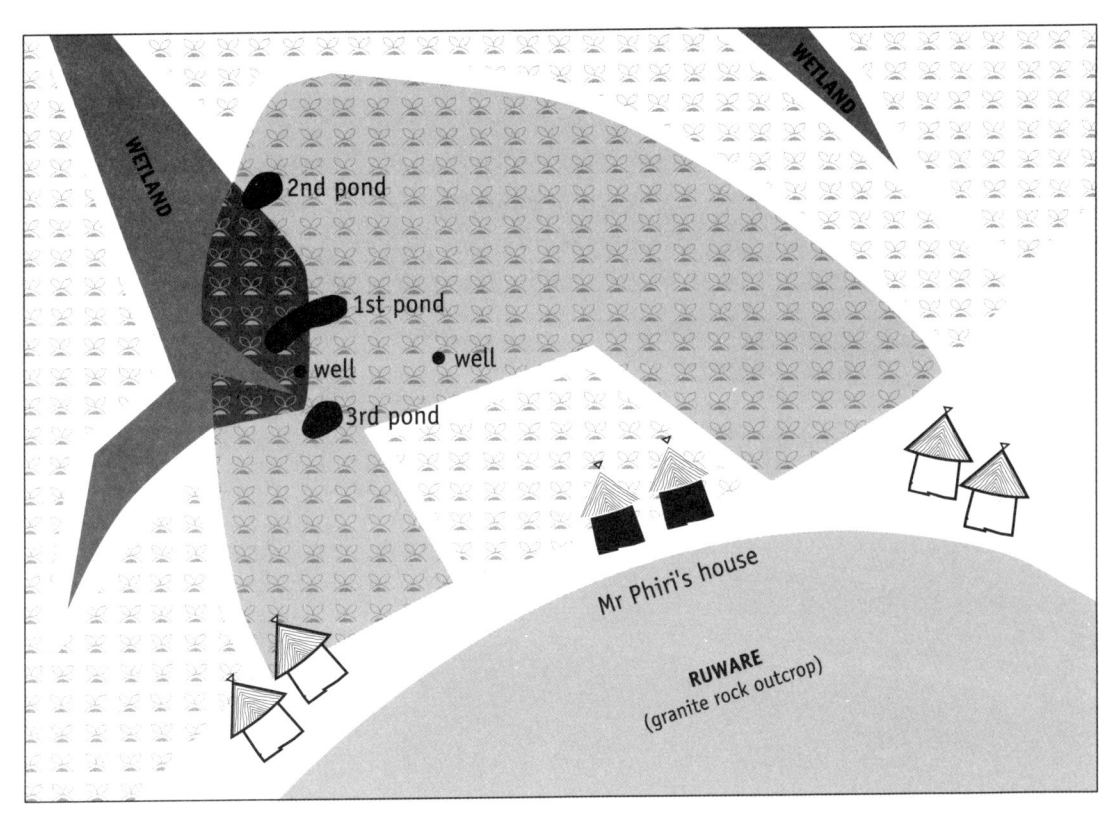

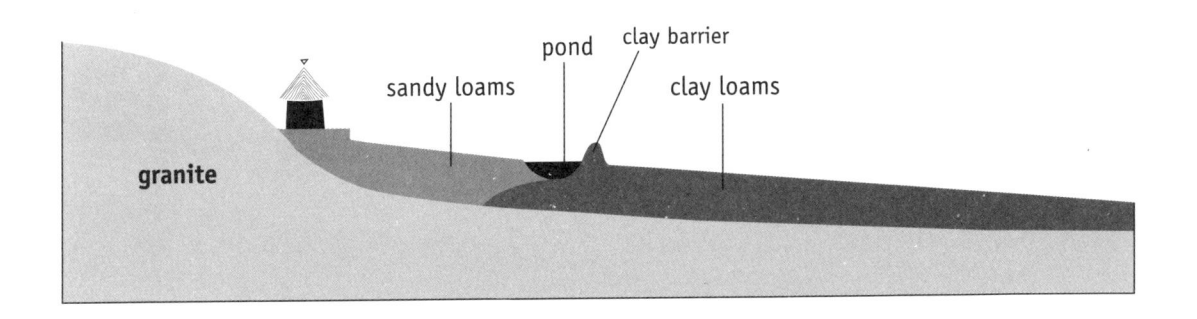

| | Mr Phiri's fields, 8 acres in total | | Neighbours' fields |

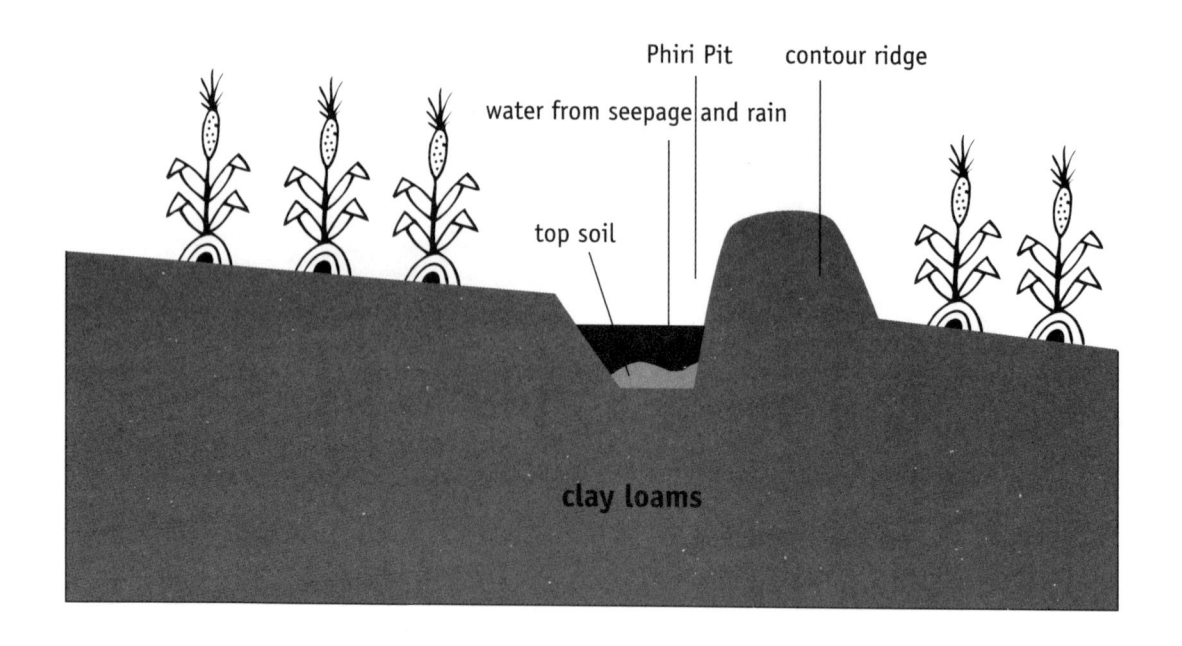

Phiri Pit   contour ridge
water from seepage and rain
top soil
clay loams

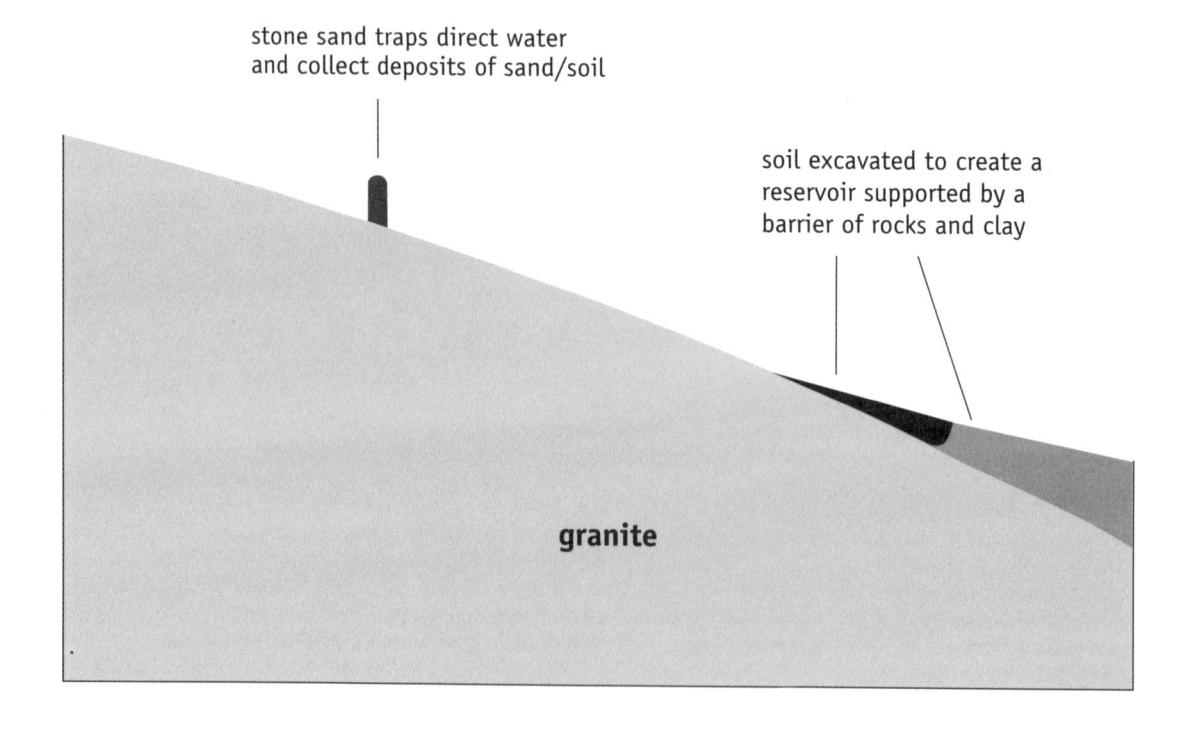

stone sand traps direct water
and collect deposits of sand/soil

soil excavated to create a
reservoir supported by a
barrier of rocks and clay

granite

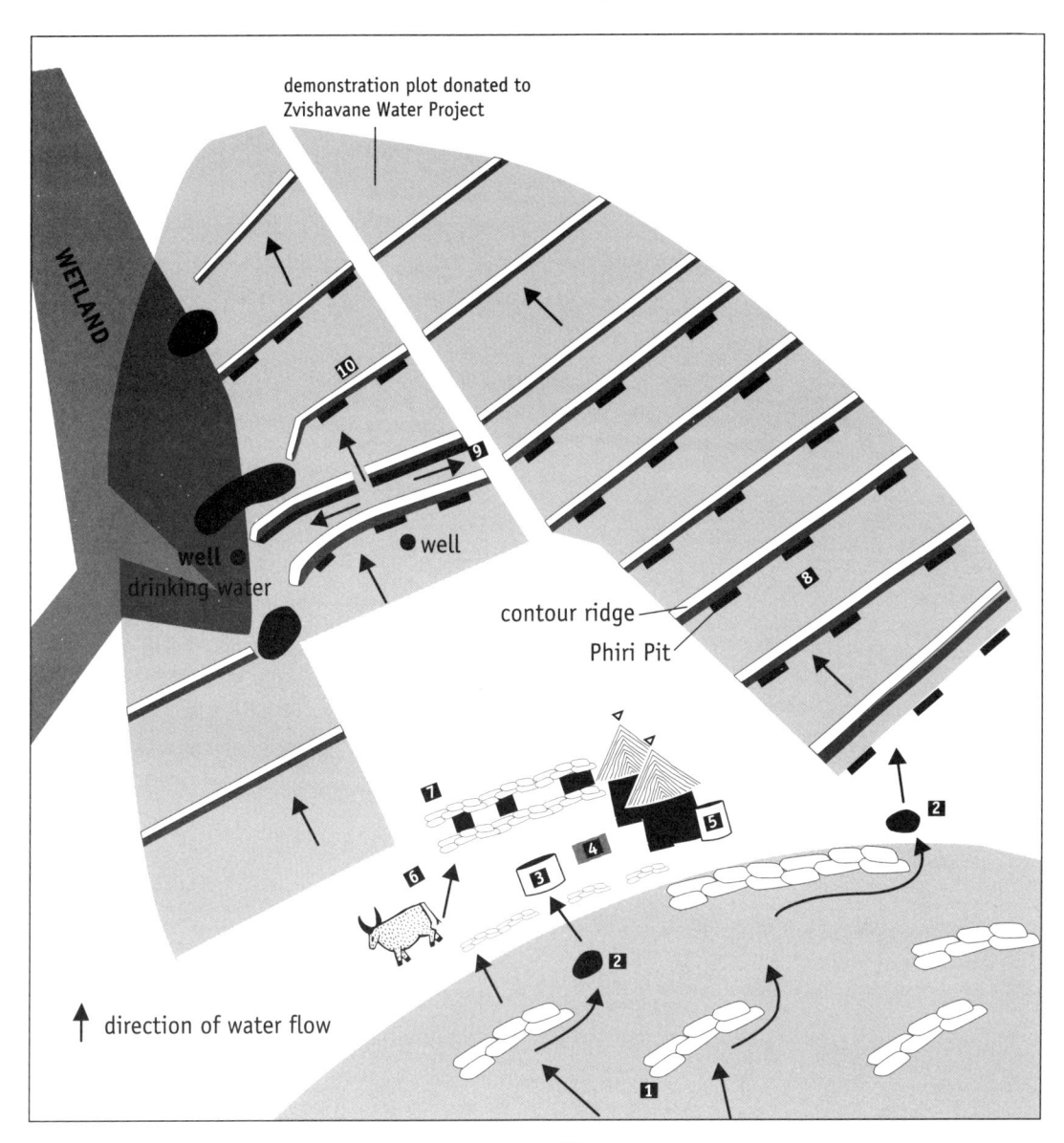

Labels on diagram:
- demonstration plot donated to Zvishavane Water Project
- WETLAND
- well drinking water
- well
- contour ridge
- Phiri Pit
- ↑ direction of water flow

1 **Sand traps** made from broken stones slow down rainfall to separate sand/soil from water and direct the water towards catchments.

2 **Hand-dug reservoirs** collect rainfall and allow water seepage through soil/fields (see illustration p.60).

3 Overflow water from the reservoir is channelled through a pipe, for storage in a cement surface tank further down the slope.

4 Dish water is channelled into the **'Poor Man's Tank'**, an underground stone-filled pit that also catches seepage.

5 The **'Demonstration Tank'** harvests rainwater from the roof and releases it by tap.

6 Rain washes manure from cattle kraal into sand traps
7 and pits where it is retrieved to fertilize fields.

8 Earthen **Phiri Pits** within contour ridges collect rainfall, water seepage and run-off soil that can be scooped out and applied to fields.

9 Behind the contour ridge, **natural tanks** trap water that can be channelled through pipes for irrigation, and in abundant rainfall years, can be used to raise fish.

10 **Stone-walled canals**, fed by pipes from the pond, irrigate crops when needed.

Contact the Zvishavane Water Project

Zvishavane Water Project
PO Box 118
Zvishavane
Zimbabwe

Telephone (263-151) 3250